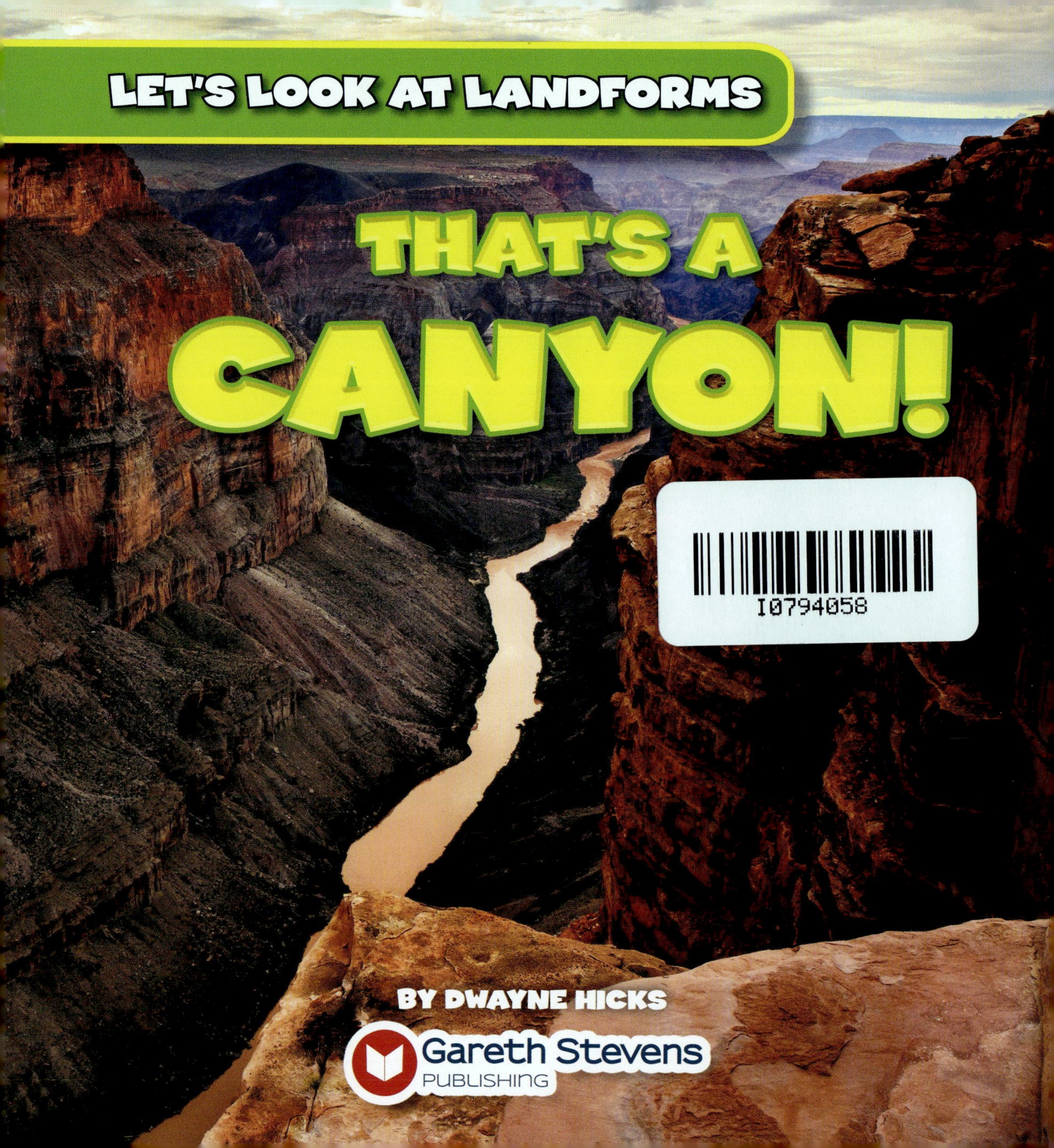

LET'S LOOK AT LANDFORMS

THAT'S A CANYON!

BY DWAYNE HICKS

Gareth Stevens PUBLISHING

Please visit our website, www.garethstevens.com. For a free color catalog of all our high-quality books, call toll free 1-800-542-2595 or fax 1-877-542-2596.

Library of Congress Cataloging-in-Publication Data
Names: Hicks, Dwayne, author.
Title: That's a canyon! / Dwayne Hicks.
Other titles: That is a canyon!
Description: New York : Gareth Stevens, [2022] | Series: Let's look at landforms | Includes index.
Identifiers: LCCN 2020032257 (print) | LCCN 2020032258 (ebook) | ISBN 9781538263679 (library binding) | ISBN 9781538263655 (paperback) | ISBN 9781538263662 (set) | ISBN 9781538263686 (ebook)
Subjects: LCSH: Canyons–Juvenile literature.
Classification: LCC GB562 .H52 2022 (print) | LCC GB562 (ebook) | DDC 551.44/2–dc23
LC record available at https://lccn.loc.gov/2020032257
LC ebook record available at https://lccn.loc.gov/2020032258

Published in 2022 by
Gareth Stevens Publishing
29 E. 21st Street
New York, NY 10010

Designer: Tanya Dellaccio
Editor: Greg Roza

Photo credits: Cover Nadia Naydenova/500px/Getty Images; pp. 3–22 (background texture) MicroOne/Shutterstock.com; p. 5 Craig Hastings/Moment/Getty Images; p. 7 Ron and Patty Thomas/E+/Getty Images; p. 9 Moritz Wolf/Getty Images; p. 11 Gergely Almási/EyeEm/Getty Images; p. 13 bradwetli photography/Moment/Getty Images; p. 15 LordRunar/E+/Getty Images; p. 17 Simon Marty/EyeEm/Getty Images; p. 19 Merrill Images/Corbis Documentary/Getty Images; p. 21 Dean Fikar/Moment/Getty Images.

Printed in the United States of America

CPSIA compliance information: Batch #CSGS22: For further information contact Gareth Stevens, New York, New York at 1-800-542-2595.

CONTENTS

Boldface words appear in the glossary.

Deep Rivers

A canyon forms when a river cuts a path into the earth. A canyon takes many years to form. Canyons have **steep** sides called cliffs. People like to spend time in and near canyons because they are so beautiful.

The Power of Erosion

Moving wind and water wear away at the earth. Tiny pieces of dirt and rock are carried away by wind and water. The pieces come to rest in a new place. This is called erosion. Erosion helps create canyons.

V Shape

A canyon starts as a river. The river wears away softer rock. Over many years, the river gets deeper. After a long time, a V-shaped canyon forms. Canyons can form on flat land. They also form in mountains.

Carried Away

River erosion makes cool shapes! Moving water and wind carry away tiny pieces of rock. These pieces are called sediment. Sediment comes to rest far **downriver**. This sediment can create new areas of land.

Cliffs and Waterfalls

Canyon cliffs often have harder rock than the land that was eroded by the river. That allows them to stand up against erosion. Some canyons have waterfalls that flow over high cliffs. They are very pretty!

A Look at the Past

A canyon's cliffs often **reveal** the past! The **layers** of colored rock in a canyon's walls show what the land was like before the river even formed. These layers piled up over many, many years. You may even find **fossils**!

People and Canyons

People have used canyon rivers to travel and ship goods for a long time. People have built **bridges** over canyons to make land travel easier. Dams on canyon rivers create large lakes called reservoirs (REH-zuh-vwahrs). Reservoir water is used for many things.

Canyon Fun!

Canyons are beautiful places! Many people love to hike in and around them. Some people love boating on canyon rivers. Other canyon activities include fishing, **tubing**, and rock climbing. Canyons are also full of awesome animals!

The Grand Canyon

The Grand Canyon is one of the biggest canyons on Earth. It's millions of years old! The Grand Canyon was formed by the Colorado River and is in the state of Arizona. It's more than one mile (1.6 km) deep in places!

GLOSSARY

bridge: a structure built over something, such as a river or canyon, so people can cross

downriver: moving toward or located near the end of a river

fossil: the marks or remains of plants and animals that formed over thousands or millions of years

layer: one part of something lying under or over another

reveal: to show clearly

steep: almost straight up and down

tubing: to use a rubber tube filled with air to float on a river or other body of water

FOR MORE INFORMATION

BOOKS

Amstutz, Lisa J. *Canyons.* North Mankato, MN: Capstone, 2020.

London, Martha. *Grand Canyon.* Minneapolis, MN: ABDO, 2020.

WEBSITES

Canyon – National Geographic

www.nationalgeographic.org/encyclopedia/canyon/

This National Geographic website discusses canyons and includes photos of famous canyons.

Fossils – Grand Canyon National Park

www.nps.gov/grca/learn/nature/fossils.htm

Learn more about the fossils discovered at the Grand Canyon from the National Park Service website.

Publisher's note to educators and parents: Our editors have carefully reviewed these websites to ensure that they are suitable for students. Many websites change frequently, however, and we cannot guarantee that a site's future contents will continue to meet our high standards of quality and educational value. Be advised that students should be closely supervised whenever they access the internet.

INDEX